1 MONTH OF
FREE
READING

at
www.ForgottenBooks.com

By purchasing this book you are
eligible for one month membership to
ForgottenBooks.com, giving you
unlimited access to our entire
collection of over 700,000 titles via
our web site and mobile apps.

To claim your free month visit:
www.forgottenbooks.com/free697682

ISBN 978-0-483-37244-3
PIBN 10697682

Laboratory Evaluation of the 1-on-10 Slope Ice Harbor Fishway Design

by Clark S. Thompson and Joseph R. Gauley

SPECIAL SCIENTIFIC REPORT–FISHERIES No. 509

UNITED STATES DEPARTMENT OF THE INTERIOR

FISH AND WILDLIFE SERVICE

BUREAU OF COMMERCIAL FISHERIES

UNITED STATES DEPARTMENT OF THE INTERIOR
Stewart L. Udall, *Secretary*
John A. Carver, Jr., *Under Secretary*
Stanley A. Cain, *Assistant Secretary for Fish and Wildlife*
FISH AND WILDLIFE SERVICE, Clarence F. Pautzke, *Commissioner*
BUREAU OF COMMERCIAL FISHERIES, Donald L. McKernan, *Director*

Laboratory Evaluation of the 1-on-10 Slope Ice Harbor Fishway Design

By

CLARK S. THOMPSON and JOSEPH R. GAULEY

United States Fish and Wildlife Service
Special Scientific Report--Fisheries No. 509

Washington, D.C.
June 1965

CONTENTS

Laboratory Evaluation of the 1-on-10 Slope Ice Harbor Fishway Design[1]

By

CLARK S. THOMPSON and JOSEPH R. GAULEY
Fishery Biologists (Research)
Bureau of Commercial Fisheries Fish-Passage Research Program
Seattle, Washington

ABSTRACT

A six-pool, full-scale section of the 1-on-10-slope Ice Harbor fishway design was built and tested in the Fisheries-Engineering Research Laboratory at Bonneville Dam before constructing the prototype. Performance of chinook salmon (Oncorhynchus tshawytscha), steelhead trout (Salmo gairdneri), and sockeye salmon (Oncorhynchus nerka) was examined in a series of tests. Results were compared with data from previous tests with a 1-on-16-slope fishway. These comparisons indicate that the new 1-on-10-slope fishway will pass fish as efficiently as the conventional 1-on-16-slope fishway.

Several modifications of the original design were examined and minor changes in it recommended. Responses of fish to various flow conditions in the test fishway also were noted and velocity profiles obtained in several typical pools.

INTRODUCTION

Ice Harbor Dam on the Snake River near Pasco, Wash., began operation in 1962. At the same time, a new fishway design was introduced. It stemmed from research by biologists of the U.S. Fish and Wildlife Service, Bureau of Commercial Fisheries, at the Fisheries-Engineering Research Laboratory, Bonneville Dam. Before constructing the prototype, a full-scale, six-pool section of the fishway was built and tested in the laboratory to determine the feasibility of departing from the standard 1-on-16 slope[2] to the more economical but steeper 1-on-10-slope Ice Harbor design.

Earlier research suggested that fishways with slopes steeper than 1-on-16 might be satisfactory for passing migrant salmonids. Gauley (1960) and Gauley and Thompson (1963) found that salmonids in a six pool, 1-on-8-slope fishway without orifices passed as fast as or faster than salmonids in a 1-on-16-slope fishway. Collins, Elling, Gauley, and Thompson (1963) further tested the two fishway slopes by comparing the performance of

individual chinook salmon (Oncorhynchus tshawytscha), sockeye salmon (O. nerka), and steelhead trout (Salmo gairdneri) in "endless" fishways. In both fishways, salmonids readily ascended 104 pools; several fish were allowed to ascend 1,000 pools or more. Generally, passage times did not differ significantly, and measurements of the degree of fatigue (based on blood lactate levels) indicated that ascent in both fishways was only a moderate exercise for salmonids when satisfactory hydraulic conditions prevailed.

From the foregoing research, the U.S. Army Corps of Engineers began to develop the new 1-on-10-slope Ice Harbor fishway design. The final design evolved from a series of miniature model tests made by the Corps at the Bonneville Hydraulic Laboratory. When suitable hydraulic conditions were established in the model fishway, the resultant design was tested under actual prototype scale operating conditions with fish. These tests were in two phases: (1) Evaluating a full-scale, six-pool test section of the fishway under laboratory conditions where provisions for change could be made before casting the prototype in concrete and (2) evaluating the prototype at Ice Harbor Dam under normal operating conditions.

A report of the laboratory studies (phase 1), conducted from April 21 to September 30, 1960, at the Fisheries-Engineering Research

[1] Financed by U.S. Army Corps of Engineers as part of a broad program of fisheries-engineering research to provide design criteria for fish-passage facilities at Corps projects on the Columbia River.
[2] Fishway slope is defined as the ratio of the rise or vertical distance to the run or horizontal distance.

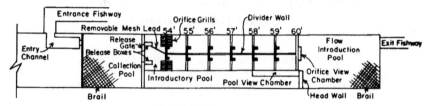

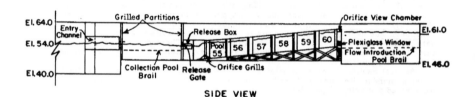

Figure 1.--Diagrammatic plan and side views of the 1-on-10-slope Ice Harbor fishway design showing temporary divider wall and other structural components in the Fisheries-Engineering Research Laboratory.

Laboratory, follows: The purposes of this experiment were to provide information on the performance of salmonids in the proposed 1-on-10-slope fishway before constructing the prototype at Ice Harbor and to develop and test optional features in the design that might further facilitate the passage of fish.

EQUIPMENT AND PROCEDURE

Laboratory

Collins and Elling (1960) fully described the Fisheries-Engineering Research Laboratory. Figure 1 shows plan and side views of the test fishway and adjacent laboratory facilities. Principal features include a collection pool, a test area, and a flow introduction pool. All units are housed in a closed wooden building where light conditions can be controlled.

Collection pool.--Fish diverted from the Washington shore fishway ascended an entrance fishway adjacent to the laboratory and entered a collection pool 28 feet long by 24 feet wide by 6 feet deep. During the tests, a wire mesh brail usually confined the fish to the upper 3 feet of the collection pool. A grill at each end of the collection pool retained the fish while permitting a flow of water through the pool.

Test area.--The test area measured 24 by 70 feet, and the experimental fishway occupied

most of this space. An introductory pool was located at the lower end of the fishway between the collection pool and weir 54^3. Release boxes located on the upstream face of the collection pool grill were used to pass individual fish and small groups of fish into the introductory pool, where they began to ascend the fishway on their own volition. Large groups of fish were released through a 5-foot-wide gate located between the two release boxes (fig. 1).

Flow introduction pool.--Water for operating the fishway was piped from the Bonneville forebay and discharged into the flow introduction pool through an adjacent diffusion chamber. Sliding gate valves controlled the amount of water entering the diffusion chamber. An exit fishway connected to the upper end of the flow introduction pool provided the fish with a return route to the main fishway. Water from this source was also used for operating the test fishway.

Lighting.--A battery of 1,000-watt mercury-vapor lamps suspended 6 feet above the water surface and 6 feet apart lighted the test fishway and adjacent facilities. Average light intensity at the water surface was about 700 foot-candles--an intensity comparable to natural lighting in the main Bonneville fishway during a typical bright cloudy day.

---

[3] All weir designations are identified in feet above mean sea level.

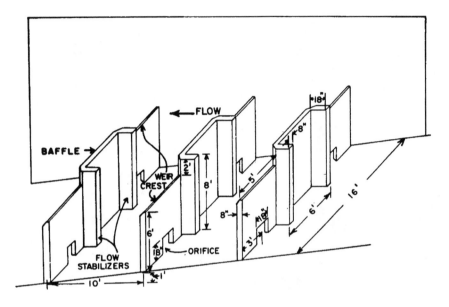

Figure 2.--Pool and weir dimensions of the Ice Harbor fishway design.

Ice Harbor Fishway Design

The new 1-on-10-slope fishway is a pool-and-overfall type with submerged orifices, flow stabilizers, and a nonoverfall section in the middle of each weir (fig. 2). There is a 1-foot rise between pools, and average water depth under normal operating conditions is 6.5 feet.

The six-pool laboratory installation was constructed of wood so that it could easily be modified to produce a variety of test conditions.

Test Conditions

Performance of fish in the Ice Harbor fishway design was compared under both normal operational conditions and experimental situations (table 1). During special tests, we modified the weir crest design and fishway hydraulics and compared the resulting fish passage under full-width and half-width fishway conditions. A narrow partition in each pool, which in effect created two 8-foot-wide fishways, produced the half-width fishway condition (figs. 1 and 3).

Underwater facilities were provided for viewing fish and fishway hydraulics.

Table 1.--Test conditions during fishway evaluation tests

Test condi- tion	Weir crest	Hydraulic condition	
		Fishway flow	Head on weirs
	Type	Type	Feet
[1] 1	The Dalles	Overfall and orifice	0.95
2	McNary	Overfall and orifice	0.95
2A	McNary	Overfall and orifice	1.20
2B	McNary	Overfall[2]	0.95
2C	McNary	Overfall[2]	1.20
3	--	Orifice[3]	0.00
4	Plane-surface ogee	Overfall and orifice	0.95

[1] Proposed normal fishway operation.
[2] Orifices closed.
[3] No overfall on weir.

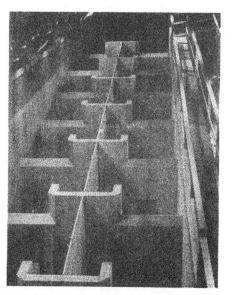

Figure 3.--Six-pool test section of the 1-on-10-slope Ice Harbor fishway design. Divider panels inserted for optional testing under a half-width condition.

FLOW

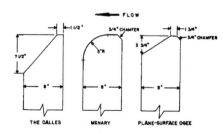

Figure 4.--Sectional views of types of weir crests.

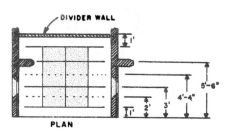

Figure 5.--Metering stations in the test fishway pool divided at center line. Dotted lines in planview indicate additional metering stations at center line (elevation) of orifices only.

Weir crest design.--Three types of weir crests were used in the tests (fig. 4). The original design specified The Dalles-type crest, but to eliminate certain undesirable characteristics of the overfall, we tested two other types of crests, the McNary and plane-surface ogee.

Hydraulic conditions.--The normal head on weirs, measured 4 feet upstream of the weir crest, was 0.95 foot, which produced a fishway flow of about 63 cubic feet per second (c.f.s.). Velocities during normal operation ranged from about 1 foot per second (f.p.s.) on the surface to about 8 f.p.s. in the center line of the orifices. Metering stations (fig. 5) indicate the points at which velocity readings were obtained in pools 57 and 58 of the half-width fishway. Velocity plots during normal flows (fig. 6) and experimental flows (figs. 7 and 8) were taken with a cup-type current meter in the left bank of the divided fishway.

Experimental flows included: (1) Overfall and orifice flows with 1.20 feet of head on weirs, (2) overfall flows only (no orifice) with 0.95 foot and 1.20 feet of head on weirs, and

(3) orifice flow only (no overfall flow) with 1.0-foot head on the orifice. When orifices were closed, 0.95-foot head on weirs produced a plunging[4] flow, but streaming[5] flows resulted when the head on the weirs was increased to 1.20 feet.

---

[4] In a plunging flow, the strong directional flow carries downward and along the bottom of the pool.
[5] A streaming flow condition produces a strong directional flow along the surface of the pool.

4

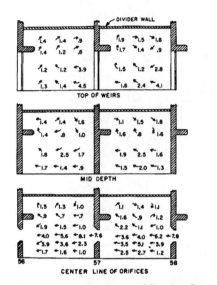

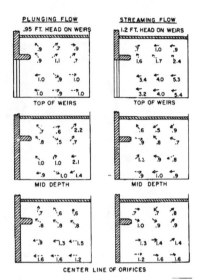

Figure 6.--Velocities (f.p.s.) and direction of flow in pools 57 and 58 during normal operation (test condition 1).

Figure 7.--Velocities (f.p.s.) and direction of flow in pool 57 during plunging and streaming flows with orifices closed. Broken lines indicate unstable flow directions.

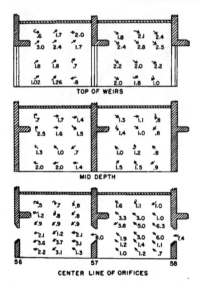

Figure 8.--Velocities (f.p.s.) and direction of flow with all of the discharge confined to the orifices.

Figure 9.--Interior of the test fishway at pool 59, showing attached observation chamber. Plexiglass wall panels are 3/4-inch thick.

Figure 10.--Observation chamber (A) for viewing passage of fish leaving the fishway through the orifices. Grillwork adjacent to window (B) deflects fish toward observer for better view and also prevents fish from drifting back through orifice after they have passed into the flow introduction pool (foreground).

Observation facilities.--Two underwater chambers were provided for observing fish and hydraulic conditions. One chamber was placed adjacent to pools 59 and 60 (fig. 9) to inspect the flow patterns and fish movement; and the other chamber at weir 60 (fig. 10), to observe fish as they passed through the orifices on completing their ascent

Figure 11.--Two release compartments and introductory area of the Ice Harbor fishway. Observer on right has opened gate to receive a fish from the collection pool. Observer on left has released a fish into the divided fishway and is recording progress of the fish as it passes over weirs in the six-pool ascent. Button switches on the handrail are used to transmit weir crossings to a time-event operations recorder.

of the fishway. Grillwork on the upstream side of each of these orifices helped to prevent fish from swimming back into the fishway from the flow introduction pool and aided in the observation of fish by deflecting them toward the viewing window.

In a model study, Corps personnel at the Bonneville Hydraulic Laboratory tested the observation chamber at the head of the fishway (weir 60) to determine its effect on fishway hydraulics. They found no noticeable effect.

Test Procedure

Passage times required to ascend the six-pool fishway were used to compare fish performance under various test conditions. Procedures for release, timing, and comparison of passage time varied slightly within different types of tests.

Types of tests.--Three types of performance tests were made: Ascents by (1) individual fish, (2) groups of mixed species, and (3) capacity-type fish concentrations, which were similar to group releases but involved larger numbers of fish[6].

Release of fish.--Individual fish were released from either of two release boxes (fig. 11), but only one fish at a time was permitted to enter the fishway. When the fishway was divided, two fish one on each side of the fishway, could be introduced simultaneously. As soon as each fish completed the ascent, another fish was released into the fishway. Species were identified and length of fish estimated as each fish passed through the release box.

Generally, the number of fish in a group test ranged from 40 to about 300. After identification, these fish were permitted to enter the test area as rapidly as possible.

---

[6] These tests were made to examine capacity-type situations; that is, the maximum number of fish (size and species considered) that could be passed through a fishway of given size and hydraulic conditions. To approach a capacity condition, it was necessary to get as many fish as possible into the fishway in the shortest period of time.

Figure 12.--Grills prevented orifice passage of fish at weir 54. All fish entering fishway passed over weir crest where they could be observed.

In capacity-type tests, fish were collected for about 48 hours prior to a test. The collection pool brail was partially raised to concentrate fish near the surface of the pool immediately before the test started. A large entry gate between the release boxes was then opened; it provided access to the fishway from the collection pool. These tests used only half of the fishway. At the upper weir, special observers identified the species.

Timing of fish.--An electrically driven operations recorder recorded the movement of fish as they ascended the fishway. Each time a fish passed a station, the observer depressed a button switch, which in turn activated a pen that scribed a mark on a time-event chart. Chart records were translated to passage times and recorded on a daily operations sheet.

Individual fish were timed as they entered the fishway at weir 54 and as they left it at weir 60. Since grills cover the orifices at weir 54 (fig. 12), each fish crossing this weir could be observed. At weir 60, observers above the weir noted fish passing the crest, and those in the observation chamber observed their passage through the orifices. A test run was completed after the fish either had ascended the six pools of the fishway or had remained in the fishway for more than 1 hour without being observed.

During group and capacity tests, counters at weirs 54 and 60 kept a time-event record of all fish entering and leaving the fishway and

also noted all "fallback activity." [7] Time intervals of the group test periods varied, depending on the willingness of fish to ascend the six-pool fishway. Tests terminated either when all fish had passed through the fishway or when only a few slow-moving individuals remained in it. Capacity tests lasted 1 hour. The entrance gate closed 30 minutes after the start of a capacity test, but observations and counting continued another 30 minutes.

Comparison of Passage Times

Both median and mean passage times were used to compare performance of individual fish under various test conditions. Median times were based on all fish tested, including those individuals that spent more than 1 hour in the fishway without completing the six-pool ascent, whereas mean passage times were based only on these fish that completed the ascent. A table of confidence intervals (Dixon and Massey, 1957) was used to test for differences between median passage times of various tests with individual fish.

For group and capacity-type releases, a statistic called "median elapsed time" was used to assess performance. This was derived by subtracting the time at which half of the

---

[7] Fish that drift or swim back over a weir or through an orifice.

8

total release had passed the lower weir (elevation 54) from the time at which half of the total entered had passed the upper weir (elevation 60). Fallbacks were accounted for by excluding the next fish following.

RESULTS

The following analysis treats performance of both individual salmonids and groups of salmonids in accordance with various test conditions (table 1) established in the laboratory version of the Ice Harbor fishway (fig. 13).

Individuals

Individual salmonids tested under proposed fishway operating conditions (test condition 1) included chinook and sockeye salmon and steelhead trout in the full-width and half-width fishway.

Chinook.--Two hundred and forty-nine chinook were timed in the full-width fishway and 253 in the half-width fishway (table 2). A comparison of median passage times (10.5 minutes and 7.1 minutes, respectively) indicated a significantly faster passage in the half-width fishway. Inspection of the mean passage times also reflected a faster ascent under the half-width fishway condition. Figure 14 presents a graphic comparison of these ascents.

Steelhead.--Passage times of individual steelhead in the full-width and half-width fish-

way were similar (table 2 and fig. 14). The median passage time for 78 individuals ascending the full-width fishway was 5.5 minutes, whereas the median passage time for similar ascents of 151 steelhead in the half-width fishway was 4.8 minutes. The difference between these times was not significant. Mean passage times for ascents during the full-width and half-width conditions were virtually the same (6.8 minutes and 6.7 minutes, respectively).

Sockeye.--The median passage time for 35 individuals ascending the full-width fishway was 2.8 minutes, while the median passage time for similar ascents of 51 sockeye in the half-width fishway was 5.0 minutes (table 2). The difference in passage times was not significant. Mean passage times were virtually the same--7.6 minutes in the full-width fishway and 7.5 minutes in the half-width fishway.

Groups

Group releases of mixed species were made in the full-width and half-width fishway (table 3). While species composition of the groups varied, either chinook salmon or steelhead predominated in individual tests. Passage times, therefore, were considered in accordance with the dominant species in a given test.

Chinook.--All chinook group releases (table 4) were made during test condition 1 (table 1). Median elapsed passage times for the four tests in the full-width fishway ranged

Figure 13.--The six-pool full-width test section of the Ice Harbor fishway design under proposed operating conditions.

9

Table 2.--Median and mean passage times of individual chinook, steelhead, and sockeye ascending six pools of the Ice Harbor fishway under full-width and half-width conditions, May 10 to September 20, 1960 (test condition 1, table 1)

Species	Fishway condition	Date	Median passage time				Mean passage time		
			Individual fish	Lower limit[1]	Median	Upper limit[1]	Individual fish	Mean[2]	Range
			Number	Minutes	Minutes	Minutes	Number	Minutes	Minutes
Chinook....	Full-width	5/10-9/23	249	8.2	10.5	12.8	224	12.8	.3-60.1
	Half-width	5/27-9/20	253	5.8	7.1 *	9.0	245	9.9	.4-54.1
Steelhead..	Full-width	5/11-9/23	78	4.4	5.5	7.0	78	6.8	.5-23.2
	Half-width	5/27-9/20	151	4.3	4.8 N.S.	6.0	147	6.7	.4-49.4
Sockeye....	Full-width	6/14-7/12	35	1.0	2.8	7.0	35	7.6	.4-35.0
	Half-width		51	2.4	5.0 N.S.	9.4	50	7.5	.5-35.0

[1] Confidence intervals for the median derived from table A-25, Dixon and Massey (1957).
[2] Based on fish completing six-pool ascent in 1-hour period.
*Significant
N.S. Not significant.

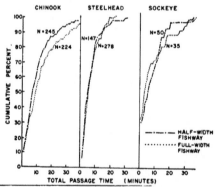

Figure 14.--Comparison of ascents of individual salmonids in the full-width and half-width fishway (test condition 1, table 1). Number of fish completing the ascent by 1-minute intervals expressed as a cumulative percent of total tested, June 8 to September 14, 1960.

from 6.6 minutes to 11.2 minutes. The average median elapsed time was 9.4 minutes. Two group releases conducted simultaneously in both sides of the divided fishway had median elapsed times of 8.8 minutes and 8.2 minutes. Results of group tests generally agree with those of tests with individual fish; that is, the fish passed faster through the half-width fishway than through the full-width fishway.

Steelhead.--Group tests of steelhead (table 4) were conducted under test condition 3 (table 1). Two group releases in the full-width fishway showed median elapsed times of 3.7 minutes and 7.2 minutes. In the half-width fishway, median elapsed times for six tests ranged from 4.7 minutes to 9.7 minutes with an average time of 6.7 minutes. Included were two tests in which fish were released simultaneously into both sides of the divided fishway. While there was a sizable range in passage times among the different tests, the data generally indicate that steelhead performed as well in the half-width fishway as in the full-width fishway.

Fishway Capacity Trials

Elling and Raymond (1959) and Elling (1960) described experiments in which large numbers of fish passed through a 1-on-16-slope fishway of pool-and-overfall design without orifices. In one of the largest releases, 50 fish per minute were passed through a 4-foot-wide fishway for 20 minutes without any indication that the capacity of the fishway had been reached.

During the experiment described here, four tests were made with a half-width section of the 1-on-10-slope Ice Harbor fishway (table 5). Performance, expressed as median elapsed time, was fairly constant during the test series, though sample sizes ranged from 460 to 1,371 fish. In June 1960, two tests, composed primarily of chinook salmon, showed median elapsed passage times of 7.1 and 7.9 minutes.

Table 3.--Species composition of group releases ascending the half-width and full-width Ice Harbor fishway, June 8 to September 14, 1960

Test condition	Date	Species composition				
		Chinook	Chinook jacks	Steel-head	Sockeye	Non-salmonids[1]
Half-width fishway		Percent	Percent	Percent	Percent	Percent
Test 1[2]...............	1	June 13	*90.0	--	7.5	2.5	--
			*90.0	--	7.5	2.5	--
Test 2.................	3	July 28	15.0	--	*77.5	1.3	6.2
Test 3.................	3	July 29	9.6	--	*87.6	--	2.8
Test 4.................	3	Aug. 3	10.4	8.8	*75.2	--	5.6
Test 5.................	3	Aug. 5	4.7	.7	*92.0	--	2.6
Test 6[2]...............	3	Aug. 12	4.9	2.5	*88.9	--	1.8
					*97.9	--	2.1
Test 7[2]...............	3	Sept. 14	12.3	3.1	*82.8	--	1.8
			23.2	1.4	*69.6	--	5.8
Full-width fishway							
Test 1.................	1	June 8	*97.6	--	--	--	2.4
Test 2-----...........	1	June 9	*94.6	--	5.4	--	--
Test 3.................	1	June 10	*91.5	--	8.5	--	--
Test 4.................	1	June 24	*77.4	2.0	7.4	13.2	--
Test 5.................	3	Aug. 9	1.0	1.4	*96.6	--	1.0
Test 6.................	3	Aug. 10	2.8	2.5	*93.3	--	1.4

[1] Includes suckers (Catostomus sp.), squawfish (Ptychocheilus oregonensis), and carp (Cyprinus carpio).
[2] Both sides of the divided fishway tested simultaneously.
*Dominant species.

These performances compare favorably with those in the group tests of June 13 (table 4) in which the median elapsed times for sample sizes of only 40 fish were 8.8 and 8.2 minutes, respectively. Capacity tests in late July and early August primarily used steelhead trout of sample sizes considerably larger than those in the June tests (table 5). Passage times in the midsummer tests are not directly comparable with those in the earlier tests because of differences in season, species composition, and weir crest design in the two sets of tests.

The general performance of fish in these tests showed that they moved freely through the fishway. The number of fish during these tests was apparently insufficient to approach a capacity condition in the fishway.

Effect of Weir Crest Design

The original Ice Harbor fishway design specified The Dalles-type weir crests, but in our studies we noted that this crest produced an air pocket beneath the nappe of the overfall. As there was some concern that this condition might impede the passage of fish, we also studied the effects of two other shapes of weir crests, the McNary type and the plane-surface-ogee type (fig. 4). We compared passage times of fish under various weir crest conditions (table 6).

The McNary crest appreciably reduced the objectionable air space and provided a smoother overfall and a less turbulent flow pattern in the pools (fig. 15). Fish passage

11

Table 4.--Comparisons of median elapsed times of group releases of chinook salmon and steelhead trout ascending the half-width and full-width Ice Harbor fishway, June 8 to September 14, 1960

	Dominant species	Test condition[1]	Date	Fish entering fishway	Fish completing fishway	Test Period	Median elapsed time
Half-width fishway				Number	Percent	Minutes	Minutes
Test 1[2].............	Chinook	1	June 13	40	92.5	85	8.8
				40	87.5		8.2
Test 2...............	Steelhead	3	July 28	75	92.0	64	4.7
Test 3...............	Steelhead	3	July 29	183	100.0	76	8.2
Test 4...............	Steelhead	3	Aug. 3	120	93.3	76	5.0
Test 5...............	Steelhead	3	Aug. 5	148	98.6	75	6.6
Test 6[2].............	Steelhead	3	Aug. 12	77	100.0	84	4.8
				48	95.8		9.7
Test 7[2].............	Steelhead	3	Sept. 14	154	99.4	104	7.2
				69	95.6		7.2
Full-width fishway							
Test 1...............	Chinook	1	June 8	41	92.7	85	11.2
Test 2...............	Chinook	1	June 9	74	94.6	103	9.2
Test 3...............	Chinook	1	June 10	47	78.7	88	10.4
Test 4...............	Chinook	1	June 24	51	98.0	56	6.6
Test 5...............	Steelhead	3	Aug. 9	200	100.0	106	3.7
Test 6...............	Steelhead	3	Aug. 10	282	97.2	88	7.2

[1] See table 1.
[2] Both sides of the divided fishway tested simultaneously.

Table 5.--Capacity tests in the 1-on-10-slope Ice Harbor fishway (half-width), 1960. Six-pool ascent, 60-minute test periods

Date	Fish entering fishway	Fish leaving fishway	Fish completing fishway	Median elapsed passage time	Species composition				Weir type
					Chinook	Steelhead	Sockeye	Non-salmonids[1]	
	Number	Number	Percent	Minutes	Percent	Percent	Percent	Percent	
June 20	460	464	100.0	7.1	72.5	3.2	19.4	4.9	The Dalles
June 27	816	804	98.1	7.9	75.5	8.3	4.0	12.2	The Dalles
July 25	1,371	1,344	98.0	9.2	19.8	77.4	1.1	1.7	McNary
Aug. 1	1,256	1,165	92.8	10.0	14.6	83.5	0.6	1.3	McNary

[1] Includes suckers (Catostomus sp.), squawfish (Ptychocheilus oregonensis), and carp (Cyprinus carpio).

Table 6.--Median and mean passage times of individual chinook, steelhead, and sockeye ascending six pools of the half-width Ice Harbor fishway under three weir crest conditions, July 13 to September 16, 1960

Type of crest and species tested	Test cond.[1]	Date	Median passage time				Mean passage time[3]		
			Individual fish Number	Lower limit[2] Minutes	Median Minutes	Upper limit[2] Minutes	Individual fish Number	Mean Minutes	Range Minutes
The Dalles crest									
Chinook............	4 1	July 13-21	27	4.3	9.2	18.0	26	11.4	1.2-35.6
	1	Sept. 6-16	114	2.1	3.4	8.4	113	8.2	.1-42.0
Steelhead..........	4 1	July 13-21	106	3.2	5.1	6.6	106	7.1	.5-32.7
	1	Sept. 6-16	109	2.8	3.6	6.1	109	6.1	.6-39.6
Sockeye............	4 1	July 13-21	15	.6	5.4	16.7	15	9.3	.3-36.7
McNary-type crest									
Chinook............	2	July 13-21	16	.5	1.6	11.0	15	6.2	.3-40.5
	2	July 22-Aug.1	7	.5	8.7	22.2	7	11.2	.5-22.2
	2	Sept. 6-16	104	1.0	1.9	5.6	103	7.1	.2-55.4
Steelhead..........	2	July 22-Aug.1	98	1.4	2.3	3.2	98	4.8	.4-40.4
	2	July 13-21	140	2.8	4.0	4.8	138	4.9	.4-30.8
	2	Sept. 6-16	120	1.8	2.4	3.4	120	4.2	.4-30.0
Sockeye............	2	July 13-21	17	.6	.9	5.1	17	4.7	.4-22.5
Plane-surface ogee crest									
Chinook............	4	July 22-Aug. 1	10	.4	3.2	9.0	10	4.0	.4-10.1
Steelhead..........	4	July 22-Aug. 1	89	2.0	2.6	3.2	89	4.9	.4-62.4

[1] See table 1.
[2] Confidence intervals for the median derived from table A-25, Dixon and Massey (1957).
[3] Based on fish completing six-pool ascent in 1-hour period.
[4] Weir baffles chamfered.

13

Figure 15.--Operational view of the Ice Harbor fishway with McNary crests on right and The Dalles crests on left. Note reduced turbulence in flow pattern on right. Corners on downstream face of the weir baffles on left are chamfered and those on right squared. Eventual design of prototype was modified to conform with structure on right.

also improved over that under The Dalles-type condition. Generally, faster ascents occurred when the McNary crest was used (table 6).

The plane-surface-ogee-type crest also effectively reduced the air space beneath the weir nappe. Performances of fish under this condition compared favorably with those under the McNary crest condition for the same general period.

The generally improved hydraulic condition and favorable performance of fish under the modified weir crest conditions led to a recommendation that McNary-type crests be used in the prototype.

Weir Baffle Design

During some of the tests the downstream face of the weir baffles were chamfered and on others they were squared (fig. 15). With a chamfered face on the weir baffles, overfall flows flared toward the center of the pool (fig. 13). Squaring the face of the weir baffles appeared to improve the weir overfall condition by confining the spill to a direct in-line flow.

Effect of Flow Conditions

The proposed Ice Harbor fishway was designed to operate with an overfall and orifice flow with about 1 foot of head on the weirs. During evaluation tests, various flow conditions (figs. 16 and 17) were examined to assess their effect on passage of individual chinook salmon and steelhead trout. These included (1) overfall and orifice flows with 0.95-foot and 1.20-feet of head on the weirs, (2) overfall flows with 0.95-foot and 1.20-feet head on the weirs and with the orifices closed, and (3) orifice flows with no overfall. McNary-type crests were installed during these tests.

Chinook.--Results of tests with individual chinook salmon (table 7) showed that certain flow conditions materially affected their movement in the fishway. Under the overfall and orifice flow condition, the median passage time was significantly less when the head on the weirs was 1.20 feet than when it was 0.95 foot. When only overfall flows prevailed and the head on the weir was 1.20 feet, the median passage time was significantly greater than that under the overfall flow condition with 0.95-foot head on the weirs.

Performance under orifice flows only (without overfall) compared favorably with that under the overfall-flow-only condition with 0.95-foot head on the weirs. When only orifice flows prevailed, chinook salmon responded quite readily to this hydraulic condition and spent little time in roaming about the pool.

14

Figure 16.--Operational view of the Ice Harbor fishway with McNary-type weir crests on both sides. Orifices are functional on the left and closed on the right. Weirs have 0.95-foot head.

Figure 17.--Operational view of the Ice Harbor fishway during orifice flow.

15

Table 7.--Median and mean passage times of individual chinook salmon ascending six pools of the half-width Ice Harbor fishway under various flow conditions, August 22 to September 30, 1960

Flow	Head on weirs	Test condi- tion[1]	Date	Median passage time				Mean passage time		
				Individual fish	Lower limit[2]	Median	Upper limit[2]	Individual fish	Mean[3]	Range
Type	Feet			Number	Minutes	Minutes	Minutes	Number	Minutes	Minutes
Overfall and orifice flow	0.95	2	8/15-9/20	44	2.1	5.6	8.1	44	7.9	.4-43.2
	1.20	2A	8/22-9/2	36	.6	.9	8.2	35	4.6	.4-27.3
Overfall flow only	0.95	2B	8/15-9/2	33	1.3	2.8	6.8	33	5.6	.5-22.0
	1.20	2C	8/22-9/2	13	5.2	9.5	19.7	13	11.2	.4-27.5
Orifice flow only	4 0.00	3	8/2-9/30	57	1.5	2.6	4.8	56	6.7	.5-42.9

[1] See table 1.
[2] Confidence intervals for the median derived from table A-25, Dixon and Massey (1957).
[3] Based on fish completing six-pool ascent in 1-hour period.
[4] The slight overfall noted on weirs 55 and 56 in figure 17 occurred when the pool surface rose during surges of water through the orifices.

16

During overfall flow conditions with 1.2 feet of head, the fish often moved about the pool considerably before passing to the next pool.

Steelhead.--The performance of steelhead under various flow conditions (table 8) was similar to that evidenced in tests with chinook salmon. The increase in head on the weir induced faster ascents when orifices were open and slower ascents when orifices were closed. The fastest ascent occurred when only orifice flows were present, and the slowest ascents occurred during the overfall-only condition.

Observations from the viewing area revealed that steelhead actively explored the lower half of the pool when orifices were closed. During test condition 2C (orifices closed and high flow), steelhead nearly always rested on tb bottom of the pool. Characteristic behavic during orifice flow as direct passage throug the pool in line with the orifices.

Weir Overfall vs. Orifice Passage

Observations were made at weir 60 to determine if various species preferred the weir overfall or the orifice during ascent. Tests were made on individual chinook, steelhead, and sockeye, and on two group releases of all species available.

The percentage of individual chinook, steelhead, and sockeye using weir overfalls and orifices was calculated for May through September (table 9). Chinook and steelhead showed a seasonal difference in behavior. During May and June most chinook preferred the orifice, but during July, August, and September they preferred the weir overfalls. Steelhead, however, preferred weir overfalls early in the season and orifices later on. Sockeye consistently preferred the weir overfalls.

Two group releases, comprised of both salmonids and other fish, provided additional information on the preference of various species for either the overfall or the orifice during passage (table 10). Salmon and steelhead responses were comparable to those in the tests with individual fish. Squawfish (Ptychocheilus oregonensis), suckers (Catostomus sp.), and carp (Cyprinus carpio) decidedly preferred orifice passage while shad (Alosa sapidissima) predominantly preferred the overfall.

COMPARISON OF PERFORMANCE IN 1-ON-10- AND 1-ON-16-SLOPE FISHWAYS

A preliminary evaluation of the test fishway may be made by comparing passage times of fish in the six-pool Ice Harbor design to similar data from previous tests in a six-pool section of a 1-on-16-slope fishway without orifices. Passage times by species (fig. 18)

are shown as a cumulative percentage of all fish completing an ascent of six pools in 60 minutes or less. Data for the 1-on-16-slope fishway apply to a structure 11.5 feet wide, whereas the 1-on-10 material is based on tests conducted in fishways both 8 and 16 feet wide. While seasonal distribution of fish in the respective comparisons differed, the major portion of the chinook, steelhead, and sockeye runs in each year was covered fairly well.

These comparisons show that the three species made faster ascents in the 1-on-10-slope fishway than in the 1-on-16-slope fishway. The difference in performance of chinook and sockeye salmon in the two fishways was less than that of steelhead, but a trend of somewhat faster ascent in the 1-on-10 slope was still apparent. _

The performance of fish in the two fishway slopes may be further compared by examining results of capacity tests in the respective fishways. Tests in a 1-on-16-slope fishway on June 25, 1957, (Elling, 1960) can be compared with the tests of June 20 and 27, 1960, in this report. Mostly chinook salmon were used in the 1957 and 1960 tests. Respective median elapsed times for the 1957 tests were 9.2, 10.0, and 13.1 minutes, whereas in the 1960 tests, passage times for a similar ascent of six pools were 7.1 and 7.9 minutes. Again, these data appear to parallel results of the comparison with individual fish; that is, a somewhat faster ascent occurred in the 1-on-10-slope fishway than in the 1-on-16-slope fishway.

The foregoing analysis must, of course, be considered with some reservation because conditions for passage were not comparable in the two fishways; that is, orifices were present in the 1-on-10 slope and lacking in the 1-on-16 slope. Without knowing how orifices would have affected passage in the 1-on-16-slope fishway, we must await comparisons of both fishway designs with orifices under prototype conditions before we can make the final analysis. Nevertheless, results of the current work appear to be sufficiently encouraging to warrant a judgment that a 1-on-10-slope fishway of the design tested should be as suitable for passage of fish as the conventional 1-on-16-slope fishways now used on the Columbia River.

SUMMARY AND CONCLUSIONS

A full-scale, six-pool section of the proposed 1-on-10-slope Ice Harbor fishway design was tested in the Fisheries-Engineering Research Laboratory at Bonneville Dam before constructing the prototype. Pools were 16 feet wide by 10 feet long (weir center to weir center). The key feature of this fishway is the special weir design consisting of a center baffle and vertical flow stabilizers, with 5-foot wide overfall sections on each side

17

Table 8.--Median and mean passage times of individual steelhead trout ascending six pools of the half-width Ice Harbor fishway under three different flow conditions, August 22 to September 30, 1960

Flow	Head on weirs	Test condition[1]	Date	Median passage time				Mean passage time		
				Individual fish	Lower limit[2]	Median	Upper limit[2]	Individual fish	Mean[3]	Range
Type	Feet			Number	Minutes	Minutes	Minutes	Number	Minutes	Minutes
Overfall and orifice flow	0.95	2	8/15-9/2	183	3.1	4.3	5.7	181	5.6	.4-23.9
	1.20	2A	8/22-9/2	116	1.8	2.9	4.4	116	5.0	.3-26.8
Overfall flow only	0.95	2B	8/15-9/2	160	5.0	5.5	6.9	159	6.6	.5-25.2
	1.20	2C	8/22-9/2	71	10.8	11.9	13.9	68	13.0	1.2-64.2
Orifice flow only	0.00	3	8/2-9/30	226	1.5	1.8	2.1	223	3.8	.4-38.9

[1] See table 1.
[2] Confidence intervals for the median derived from table A-25, Dixon and Massey (1957).
[3] Based on fish completing six-pool ascent in 1-hour period.

Table 9.--Percent of individual chinook, steelhead, and sockeye using weir overfalls and orifices at weir 60, May to September 1960

Month	Chinook			Steelhead			Sockeye		
	Weir 60 count	Overfall passage	Orifice passage	Weir 60 count	Overfall passage	Orifice passage	Weir 60 count	Overfall passage	Orifice passage
	Number	Percent	Percent	Number	Percent	Percent	Number	Percent	Percent
May..........	61	39.3	60.7	41	75.6	24.4			
June..........	274	36.5	63.5	515	40.0	60.0			
July..........	107	69.2	30.8	376	40.4	59.6	53	88.7	11.3
August..........	30	60.0	40.0	340	34.1	65.9	64	84.4	15.6
September..........	293	64.2	35.8						
Total..........	765	52.8	47.2	1,272	39.7	60.3	117	86.3	13.7

18

Table 10.--Percent of different species using weir overfalls and orifices at weir 60 during two group releases, June 29 and July 6, 1960

Species	Weir 60 count	Crest	Orifice
	Number	Percent	Percent
Chinook........	29	27.6	72.4
Steelhead......	35	40.0	60.0
Sockeye........	39	94.9	5.1
Shad...........	15	73.3	26.7
Squawfish......	24	16.7	83.3
Sucker.........	233	11.6	88.4
Carp...........	8	0.0	100.0

of the fishway. There are two 18-inch square orifices in each weir located on the floor and adjacent to the sides of the fishway.

Passage times of chinook salmon, sockeye salmon, and steelhead trout were used to assess performance of fish under various operational conditions in the test fishway section. A divider panel, inserted in the center of each pool, permitted simultaneous testing of the effects of various fishway modifications on the performance of fish. Comparisons of the performance of fish under half-width (8 feet wide) and full-width fishway (16 feet wide) conditions were also made. Results of these tests, when compared with similar data from previous tests in a 1-on-16-slope fishway,

provided a preliminary basis for evaluating the new fishway design.

Flow patterns and velocity profiles were obtained under normal and experimental operating conditions. These showed velocities ranging from approximately 1 to 8 feet per second in various parts of the pool, with the majority of the flows ranging between 1 and 2 feet per second.

A summary of observations during evaluation of various conditions follows:

1. Comparisons of passage times of fish under half-width and full-width fishway conditions showed that individual chinook salmon made significantly faster ascents under the half-width fishway condition than under the full-width condition. Steelhead trout and sockeye salmon performed about the same under either condition. Tests with groups of fish indicated similar results.

2. During four capacity tests in the half-width fishway, as many as 1,371 fish entered the test area during a 30-minute entry period. Performance during these tests did not indicate that fish movement through the fishway was impeded. More than 90 percent of the fish in all tests passed through the fishway during the 60-minute observation periods.

3. Use of a McNary-type weir crest in lieu of a Dalles-type crest, specified in the original design, improved hydraulic conditions in the fishway and appeared to hasten slightly the passage of fish. Brief examination of the effects of a plane-surface-ogee crest indicated

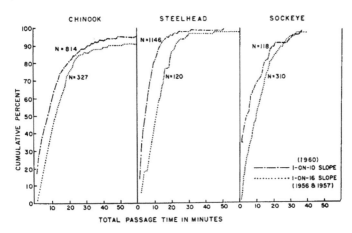

Figure 18.--Cumulative distribution of passage times of individual chinook, steelhead, and sockeye in 1-on-10- and 1-on-16-slope fishways. Number of fish completing a six-pool ascent by 1-minute intervals expressed as a cumulative percentage of total tested (1-on-10 slope--1960, 1-on-16 slope--1956 and 1957).

19

results similar to those shown when the McNary-type crest was used.

4. A minor change in the original design of the weir baffles (squared corners rather than chamfered corners on downstream face) improved the weir overfall condition.

5. Observations on the effect of various flow conditions on passage of fish showed that both chinook salmon and steelhead trout made significantly faster ascents under an overfall and orifice flow condition when the head on the weir was increased from 0.95 foot to 1.20 feet. With the orifices closed, both species made significantly slower ascents when the head on the weirs was increased from 0.95 foot to 1.20 feet. When only orifice flows were provided, steelhead ascended the fishway in less time than under any of the other conditions tested. Chinook salmon appeared to accept quite readily the orifice-flow-only condition and ascended the fishway without apparent difficulty.

6. The preference of various species of fish for orifices or overfalls during ascent was examined at the uppermost weir of the fishway. Chinook salmon preferred orifice passage during the early part of the season and reversed this preference as the season progressed. Early runs of steelhead preferred the overfall and later runs the orifice. Generally, preference ratios were about 60 to 40. Over 85 percent of sockeye salmon used the overfall passage during ascent.

Among other fish observed were carp, shad, squawfish, and suckers. Eighty-three to one hundred percent of the carp, squawfish, and suckers used the orifice passage. Shad favored the overfall.

Results of performance of salmonids in the new 1-on-10-slope Ice Harbor fishway were compared with similar data from previous tests in a 1-on-16-slope fishway. Comparisons indicate that a 1-on-10-slope fishway of the design tested is as suitable for passage of fish as conventional 1-on-16-slope fishways now used on the Columbia River.

LITERATURE CITED

COLLINS, GERALD B., and CARL H. ELLING.
1960. Fishway research at the Fisheries-Engineering Research Laboratory. U.S. Fish and Wildlife Service, Circular 98, 17 p.

COLLINS, GERALD B., CARL H. ELLING, JOSEPH R. GAULEY, and CLARK S. THOMPSON.
1963. Effect of fishway slope on performance and biochemistry of salmonids. U.S. Fish and Wildlife Service, Fishery Bulletin, vol. 63, no. 1, p. 221-253.

DIXON, WILFRID J., and FRANK J. MASSEY, JR.
1957. Introduction to statistical analysis. 2d ed. McGraw-Hill Book Company, New York, 48 p.

ELLING, CARL H.
1960. Further experiments in fishway capacity, 1957. U.S. Fish and Wildlife Service, Special Scientific Report--Fisheries No. 340, 16 p.

ELLING, CARL H., and HOWARD L. RAYMOND.
1959. Fishway capacity experiment, 1956. U.S. Fish and Wildlife Service, Special Scientific Report--Fisheries No. 299, 26 p.

GAULEY, JOSEPH R.
1960. Effect of fishway slope on rate of passage of salmonids. U.S. Fish and Wildlife Service, Special Scientific Report--Fisheries No. 350, 23 p.

GAULEY, JOSEPH R., and CLARK S. THOMPSON.
1963. Further studies on fishway slope and its effect on rate of passage of salmonids. U.S. Fish and Wildlife Service, Fishery Bulletin, vol. 63, No. 1, p. 45-62.

MS. #1406

UNITED STATES
DEPARTMENT OF THE INTERIOR
FISH AND WILDLIFE SERVICE
BUREAU OF COMMERCIAL FISHERIES
WASHINGTON, D.C. 20240

POSTAGE AND FEES PAID
U.S. DEPARTMENT OF·THE INTERIOR

Librarian,

Marine Biological Lab.,

SSR 7⁻   Woods Hole, Mass.